Money Lit Curriculum

Many thanks to the Plutus Foundation for making the creation of this curriculum, and the companion activity book, possible. The creation of this curriculum has been almost two years in the making as I have taken time to build the topics, the supporting activities and projects, and work with an illustrator to create images and the companion activity book. This curriculum was created to be an easy way for educators and to implement economic and everyday money life skills into their social studies, math, or personal finance classes. It was created with homeschooling parents and teachers in mind, but is able to be used in a traditional classroom setting as well.

I would also like to thank Kelly McKibbin for creating the images you see throughout this book.

Money Lit Lesson Plan: Month 1

HUMAN CAPITAL

In order to learn the needed money life skills, you have to start at the base of our earning potential. **Human Capital is a combination of our skills, education, and experience we develop over a lifetime that adds to the value we bring to a job and our community.** Whether we work for ourselves as entrepreneurs, work for others as an employee, or stay at home with our children, we all need to build our human capital.

Key Terms

Human Capital: Our skills, education, and experience we gain over a lifetime, that adds to our earning potential and value to our community.

Job Specific Skills: Skills and education that you need for a specific job. (Ex. A nurse needs to know the skills to perform their job and have education in the nursing field.

Soft Job Skills: Skills that you need at any job that aren't job specific, but are equally important. (Ex. Being on time, working well with others, being organized, managing your time).

MONTH #1: HUMAN CAPITAL

Human Capital: Our skills, education, and experience that we gain throughout a lifetime. We use our human capital to earn money and contribute to our communities and families.

Discussion Questions

Write your answers on a separate piece of paper

-Why do we have to go to school?

-What ways are you learning skills outside of school? Music, sports, crafting, writing, etc?

-When you do activities outside of school, what other skills do you think you are learning? (Hint: time management, working well with others)

-How do you want to earn money when you grow up? What skills and education does that require?

-Have you volunteered anywhere? What did you enjoy about that? What did you learn?

Activities

- Complete the human capital activity pages in the activity book.
- Complete the "What other skills do I need" page in the activity book.
- Using the occupation cards, answer the following questions:
    1. What skills are needed for each job type?
    2. What education is needed for each job type?

Critical thinking: Look up an occupation that isn't on this list and answer the same questions.

Human Resources
$59,180

Revenue Manager
$65,000

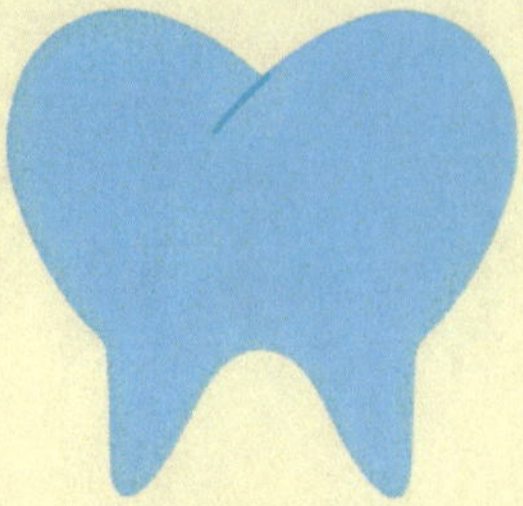

Dentist
$159,200

Graphic Designer
$76,137

Welder
$41,380

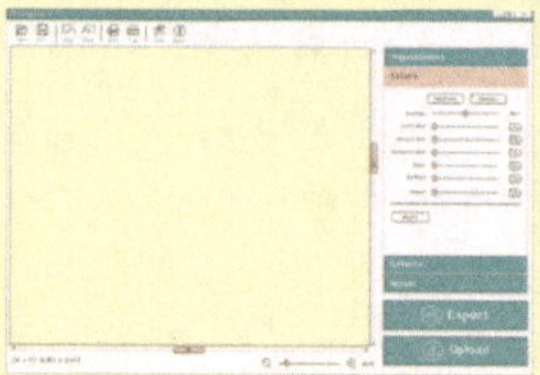

Software Developer
$70,000

Coder
$86,550

Teacher
$60,477

Building Inspector
$55,960

Restaurant Manager
$55,340

Barber
$30,480

Architect
$80,750

Electrician
$49,260

Warehouse
$28,677

Engineer
$91,010

Veterinarian
$95,460

Project
What Can I Do With My Human Capital

Materials needed: Poster board, Index Cards, pencil, glue stick

- Take three index cards and label with skills, one education, and one experience. Fill them out based on skills, education, and experience you currently have. Give them each some space on your poster board and glue them.
- Write down on another index card one or two skills you want to learn that you don't already know. Glue this to your board.
- You may decorate your board with pictures or drawings of what you think you could do to earn money, or of your current skills, education, and experience.
- Once completed, present this to your teacher/parent/guardian

Educator's Corner:

During the discussion questions, feel free to add your own that are meaningful to your students. These questions are meant to guide open discussions and are a starting point.

The project is meant to bring all the learning points together. At this point they will have said it, seen it, and wrote it more than once. If you want your student(s) to work on typing skills, they can type instead of write for step #2, and then print it to attach to the board. By having them present it, they are also practicing their public speaking and presentation skills

Skills Learned

- Introduction to an economics basic concept
- Group discussion
- Critical thinking
-Writing/typing skills
-Research skills
-Presentation skills
-Art/creativity

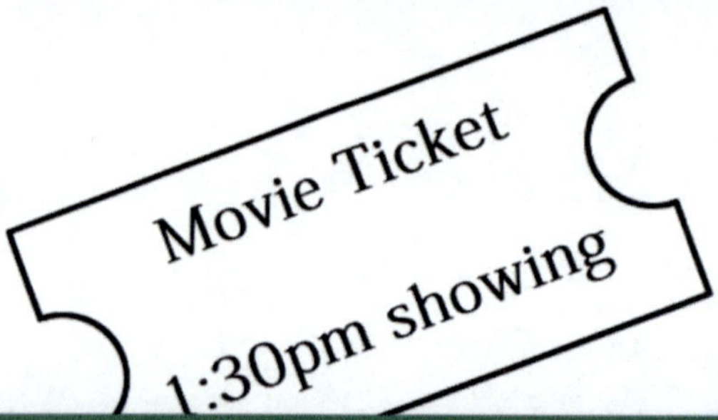

Money Lit Lesson Plan: Month 2

GOODS AND SERVICES

Recap: Last month we discussed human capital and how important it is to build our skills, get an education, and have experience. You can gain experience through volunteering, an internship with a company that does what you think you might be interested in, or through a job in the field you want to be in.

This month: We are discussing goods and services and how we use our human capital to create them.

Key Terms

Goods: Physical items we can buy with our earned money such as food, a car, a toy, a cell phone. Goods are most often referred to as *products.*

Services: A skill that someone has that we pay for such as a hair cut, car wash, lawn service, or a tutor.

Entrepreneur: Someone who sees a need and uses their human capital skills to create a product or service to fill that need, or improve on a current product or service.

Revenue: The total amount of income you make from selling your goods or services, often referred to as *gross income*

Profit: Income after the expenses, costs, and taxes are taken from the total revenue or gross income amount. Often referred to as *net income.*

MONTH #2: GOODS AND SERVICES

Discussion Questions

Write your answers on a separate piece of paper

Companies can offer their goods and services online or in a store. When you offer something in a physical store, it is called "brick and mortar". When you sell goods and services at a physical location, you will pay rent, electric, maybe water depending on your business, as well as paying employees to work at your store. You will need to pay for your products to be created, and keep enough of your products in your store so people have options to choose from. When you sell online, you pay for the website, the cost of manufacturing the goods you offer, or the cost of offering your service.

When you sell a good or service, your company brings in money (revenue). You then have to subtract the cost of the materials to make the product you sold, the cost of the place you sold it from (online or in person), and any costs you had to tell people about what you are selling, the taxes, and any other expenses your company has. What is left over is called profit. This is what you have actually made after selling your good or service.

For example, let's say you sell lemonade. You make lemonade and sell it on a hot day to people driving by. You sell it for $1 a cup. You didn't actually make $1. You have to factor in the cost of the lemons and sugar (or lemonade mix) you bought to make the lemonade, as well as the cost of the cups you are serving it in. Did you have to pay for the poster board or markers to make the sign? That is also an expense.

Profit = revenue - cost

- What are some goods (products) you and your family buy often?
- Do you buy most of them online or in person at the store?
- What are some services you and your family use often?
- Do you buy these goods and services at the same places each time? Why or why not?
- Critical Thinking: Which option for selling goods and services is more profitable (makes more money) and why?

Activities

- Complete the Economic Terms page in the activity book.

- Watch the following episodes of Shark Tank featuring kids who started their own businesses.

  - Mo's Bows, 11 yr old - bow ties

  - Jack's Stand an Marketplace, 10 yr old- lemonade/market stand.

- BabyToon, 10 yr old- baby spoon made from her design/formula

- Bee Sweet Lemonade, 10 yr old- lemonade that is healthy, tastes good, and saves the bees

Project

Where Are They Now?

Materials Needed: Computer for researching. Pencil and paper (you may also choose to type this).

After watching the clips on the goods and services the kid entrepreneurs created, we are now going to answer some questions and do some research and see where they are now.

- Take out 4 pieces of paper, or create four sections if you are typing. Write down the following questions for each paper or section. Re-watch each clip and answer the questions for each company.
- Are they selling a good or a service?
- What are they selling?
- How much does it cost?
- How much does it cost to make?
- What was their goal for going on Shark Tank?
- How much money are they asking for?
- What expenses are they going to use the money for?
- Were they offered a deal by one or more of the Sharks?
- Is their good or service offered online or in a store (brick and mortar)

Look up the companies and let's see where they are now since Shark Tank

- Are they still in business?
- Where do the sell their goods or services now?
- How much are they selling it for now? Is it more or less than what they were selling it for before going on Shark Tank? What is the price difference?

Type up your answers in a short paragraph or two answering these questions. Give a verbal presentation to your teacher.

Educator's Corner:

During the discussion portion, point out every day goods and services you and your family utilize daily. You can also use this opportunity to connect shopping local versus shopping chain stores, and the value of each option.

During the Shark Tank project, depending on the age of your child, you can also use this opportunity to discuss globalization and how companies have the option of using manufacturers in the country you live in, or can send it to another country to be made, and why they choose to do so. You can also choose to point out that when you offer goods and services online, anyone from any country all over the world is able to purchase and participate.

*If you would prefer your students read about kidpreneurs instead of watching a clip, the book "Dream, Dare, Do: A Guide to Exploring the Kidpreneur in Us All" is full of kidpreneur stories.

Skills Learned

- Introduction to an economics basic concept
- Group discussion
- Critical thinking
-Writing/typing skills
-Research skills
-Presentation skills
-Art/creativity
-Note taking/organization
-Everyday math

Money Lit Lesson Plan: Month 3

EARNING INCOME (MONEY)

Recap: Last month we spoke about how people use their human capital to produce goods and offer services to the community in exchange for revenue (money earned from selling something).
This month: We are discussing why we need earn money, and how our human capital is connected to our earning potential.

Key Terms

Income: Money received in exchange for work

Entrepreneur: Someone who sees a need and uses their human capital skills to create a product or service to fill that need, or improve on a current product or service.

Revenue: The total amount of income you make from selling your goods or services, often referred to as *gross income*

Memory Check-
What are goods and services?
What is human capital?

MONTH #3: EARNING INCOME (MONEY)

Discussion Questions

Write your answers on a separate piece of paper

-Why would people want to use their human capital to create a good or service to offer in the community in exchange for revenue (money)?

-Do we have to have money? If so why?

-What happens if we don't earn money?

-Do you earn money right now? What do you do to earn it? (Chores, walking dog, summer job, etc)

-Money is a way to buy goods or services that we need or want to have. It is an *exchange* of something we earn, for something we want or need. When we exchange money for a good or service, we give our money to a company so they can earn revenue.

-Do you remember what revenue means? Why would a company want to earn more revenue?

*Critical thinking question: When we exchange money for goods or services, we are helping companies earn revenue. This means we are supporting them with our earned dollars. Would earning more revenue and supporting a business mean more to a larger business or a smaller business? Is there a way you can support a smaller business this month?

-How do we earn money?

-You can earn money by using your human capital (skills, education, experience) to get a job that creates or sells a good. (ex, a store)

-You can get use your human capital to get a job that offers a service to others. (ex, a car wash)

-You can use your human capital to be an entrepreneur. (ex. start your own dog walking/bathing company)

Activities

- Look back in your activity book to the first page of Human Capital

- Have you added any skills, education, or experience since you completed those worksheets?

- What human capital do you have, that you can use to earn money?

- For example: If you are good with animals, maybe you could walk dogs, give dogs a bath, dog sit. If you are good at art or making things, maybe you find a place you can sell your art or goods you make.

- Complete the following worksheets in your activity book: Money Tree, Money coloring page, Build Your Own Booth,

~~Project~~ Field trip

Connecting My Human Capital with My Earning Potential

Local to Chattanooga Only.

1. Call the TN Small Business Development Center at 423-752-4301 (located at 100 Cherokee Blvd #102, Chattanooga, TN 37405). Ask for a tour and to speak to someone about connecting your skills with some entrepreneurial ideas.
2. Call Co.Lab at 423-648-2195 (located at 1100 Market Street, Suite 100, Chattanooga, TN 37402). They are a startup accelerator that supports entrepreneurial growth in Chattanooga and surrounding areas. Ask them if you can meet with someone and run

If you are not local to Chattanooga, Look up the business accelerator closest to you, or the entrepreneurial/small business help center and schedule a tour with them. If you don't have one, you can reach out to one that isn't in your area, and see if they will do a Zoom call with you and your students.

Does this mean you will always be an entrepreneur? It doesn’t have to. So why do it? You learn so many things when you get creative and do research and start a company. Even if you don’t do this as a lifetime career, you will learn so many valuable things and may even pay for your college or trade school degrees, first car, first home, or first investment portfolio. Developing an innovative perspective will serve you well in any field you go into. The opportunities are endless!

Educator's Corner:

Help encourage the students to find ways to bridge their human capital with ways to earn money. Make sure it is something they have an interest in, if not this lesson will fall short. Have them think of ways their skills can either find a product or service that is lacking in your area that they can solve, or ways they can improve on a current product or service. For example, if there is already someone in their neighborhood that mows lawns, help them problem solve a way to stand out from their potential competition should they insist on doing that to earn money.

If you want to expound on your student's writing or typing skills, you can have them write or type their notes from the in person or virtual field trip. You can have them present it as well if you are working on presentation/public speaking skills. Before the field trip, have them come up with three or four questions they want to ask. More questions will probably come up on the tour, but this ensures their specific questions are answered.

Skills Learned

- Bridging the gap between human capital and earning money
- Group discussion
- Critical thinking
- Writing/typing skills (optional)
- Research skills
- Presentation skills
- Art/creativity
- Note taking/organization
- Innovative thinking

Money Lit Lesson Plan: Month 4

ENTREPRENEURSHIP

Recap: Last month we discussed how we turn our human capital skills into finding ways to earn money.
This month: We are discussing entrepreneurship and how we can use that innovative thought process to either own our own businesses, or in the workplace.

Key Terms

Entrepreneur: Someone who sees a need and uses their human capital skills to create a product or service to fill that need, or improve on a current product or service.

Employee: A person who works for someone in exchange for a steady income.

Small Business: A personally owned company that employs less than 50 people.

Business Plan: A specific written plan of your goals for your business, and how you will reach them.

Branding: Designing your business in a way that customers recognize your company

Memory Check-

What are goods and services? *What is revenue?*
What is human capital? *What is profit?*

MONTH #4: ENTREPRENEURSHIP

Discussion Questions

Write your answers on a separate piece of paper

-What is an entrepreneur? It is someone who sees a need for a new product or service, or sees a way to better a current one. It is someone who uses innovative thinking to solve problems. You can have this innovative way of thinking when you work for someone too.

The first step to starting a business is to set goals and have a strategic plan for making money. This is called a business plan. A business plan has lots of categories, but we will focus on the following to keep it simple: Summary of your idea, marketing, and finances.

The summary of your idea would sound something like this: "I am starting a business because I want to earn money for financial goals I have set for myself. I have a short-term goal of buying a Lego set, and a long-term goal for saving for my first car. I hope to make at least 100 in the first 3 months. I will do this by: (state what you will be doing to earn money) and pricing them at: (cost per good or service offered). I hope to sell: (number of goods or services) at (cost) in order to reach this goal." The rest of your business plan will be laying out the strategic way you will market and price your goods or services so you can make a profit.

Next your company will need a name. The name should reflect the goods or services you plan to sell. You want what you offer to the customer to be clearly understood.

You will need a logo for your company. This is another part of branding your company so your customers recognize who you are and what you offer. For help with this, think of other larger brands and how they have become known for their goods/services. If you are driving down the road and you see the golden arches, you know right away that is McDonald's, even if it didn't have the word McDonald's on it, you would recognize their company. Colors matter too. If you are starting a company that focuses on the environment, you would want to choose colors that are earthy so they make sense for your brand.

Next you will need to think of how to tell others about what you are selling. Who will be your customer? Who is most likely to buy what you are selling? Where will you find your customer?

Once you've decided who your customer is, you have to decide how they find out about things they want to buy, so you know how to market to them. Do they find things out on social media? Would a flyer at the local coffee shop, hair salon, or other local place work? Do your customers live in your neighborhood? If so would a poster on your lawn work? Maybe a flyer at church? Maybe "word of mouth" marketing (simply telling people who will tell other people) is best? If you are unsure, you can always survey your potential customers and ask them how they hear about most of the products and services they buy.

Activities

- Sign up for a free Canva account (Canva.com). This is a graphic design website that allows you to design marketing items such as a logo. In the search bar, type "logo". Find a template that has a style you like, you can change out the colors to fit your brand. Plug in your company name, and change the colors to your brand colors of choice. You can use Canva to create flyers and posters for marketing. You simply download your logo, search for "flyers" or "posters", change the colors and wording to match what you are selling, and upload your logo to the flyer or poster you have created. You can print at home or Canva will print professionally for a small fee and ship you the items.

- Complete the business plan, and create your own logo page in your activity book.

Project

Entrepreneur Skills

Now that you have completed your business plan and some marketing materials. We will put it all together to create a business card. They are something small you can carry in a wallet or pocket to hand out to potential customers when the opportunity arises.

Using Canva, type "business card" into the search bar. Find a template that matches your logo and other marketing materials. It's important to keep the same style and colors throughout your logo and marketing materials so your brand is firmly established and easily recognizable. Once you've found a template that works, begin switching out the colors and information. Be sure to include your company name, service or goods you offer, and a way they can contact you, either phone number, email, or both.

Educator's Corner:

Learning this entrepreneurial process of starting a business teaches several concepts your students can use throughout life. They will learn some basic graphic design and attention to details as they create a brand. They will have to think strategically as they decide on colors, style that matches their goods or services, and marketing materials. You can direct them to look up and research what different colors are used for different business opportunities. They will also learn basic marketing as they think through ways to word flyers or posters.

They will also learn some sales experience as they pitch to others. A good way to elaborate on this is to have them come up with an elevator pitch where they can explain their business in 30 seconds to 1 minute's time. If you have a shy student, they can practice this so it will help them become more comfortable with telling potential customers what they do. There are so many more ways to expound on this month's curriculum if you choose to do so. Look for your city or countie's local resources for small businesses to assist with this endeavor.

Skills Learned

- Group discussion
- Critical thinking
- Writing/typing skills
- Research skills
- Public speaking skills
- Art/creativity
- Note taking/organization
- Innovative thinking
- Introduction to graphic design concepts
- Sales concepts
- Introduction to marketing concepts
- Goal setting for businesses

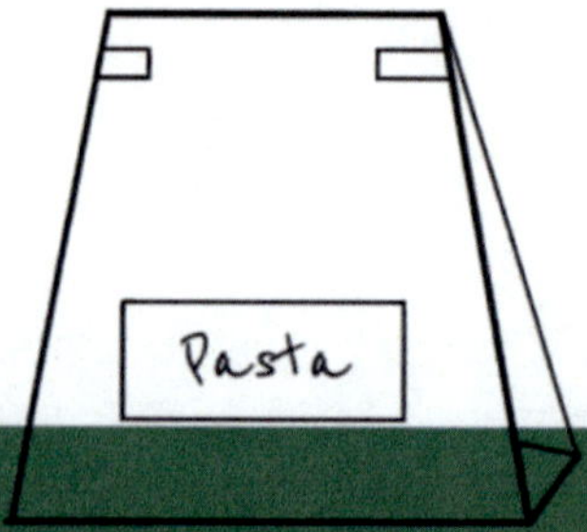

Money Lit Lesson Plan: Month 5

NEEDS VERSUS WANTS

Recap: Last month we discussed how you can use your human capital skills to become an entrepreneur. You learned the basics of business and completed a business plan.

This month: We are discussing one important tool when deciding how to spend your earned dollars, the difference between the things we need to have, and the things we want to have.

Key Terms

Need: Things we have to have for our daily survival (ex, food, water, shelter, clothing)

Want: Things we don't need for our daily survival. Things we can spend our money on *after* our needs are taken care of first.

Goal: Something you decide that you want but can't have right now so you set a plan in place to work hard for it.

Memory Check-

What are goods and services? *What is revenue?* *What is an entrepreneur?*

What is human capital? *What is profit?* *What is a business plan?*

MONTH #5: NEEDS VERSUS WANTS

Discussion Questions

Write your answers on a separate piece of paper

-We are going to start today with a fun activity. For this you will need a pencil and piece of paper. Take five minutes to think of and write down all the things you want. You can fill both sides of the paper if you want. It can be as silly as wanting your own island or airplane, or as small as wanting a pack of gum. This is your moment to dream and be silly.

You may have filled both sides of your piece of paper, or you may have written down just a few things. As we get older, the list of things we want or know are available to want grows. So why can't we just go out and buy all the things we want? We have to first understand that money is considered a limited resource. We only have as much of it as we can earn. Once we spend it on something, we don't have anymore of it until we make more. Since it is a limited resource, we have to make choices about what we should spend it on.

When we earn money, we should spend it first on things that we need. So let's make a list of things we need. On another piece of paper, write down things you *need* to live every day.

If you wrote down that our basic needs are: food, water, shelter, transportation, and clothing then you are correct. Without these we don't feel safe or well enough to do anything else, including earning money. When someone earns money from working, they should ***first*** spend their money on these basic needs before buying any wants. For adults this means we have to spend our money first on rent or mortgage payments, food, transportation, and make sure we have the weather and work appropriate clothing and shoes we need. After we take care of those things, we put money into savings for things we want or need later, then we can spend money on our wants.

Each of the basic needs listed can turn into wants if we aren't careful. It is how we spend our money that makes a difference. For example, we need food to live. Where it can become a want instead of a need is in how we spend our money on food. It is less expensive to make spaghetti at home, than to order it from our favorite Italian restaurant. That doesn't mean we can't eat out, it means we need to make wise choices with our money and if eating out every night isn't something you can fit in your budget and still afford your other needs, then it isn't the wisest use of your limited money resource.

We need clothing to survive, however clothing from consignment, thrift stores, shopping the sale racks at a department store will provide the clothing we need to cover our bodies. Buying clothes from the mall or expensive online clothing is more of a want and something we can buy once our other basic needs are taken care of. A goal can be set to save for these additional items.

We need shelter to be in a safe place and sheltered from the people and elements outside. The size house and accessories we choose to put in a house to accommodate our essential needs, versus owning a larger home we may not need or can't afford, is what determines want versus need for shelter.

Activities

-Using an old magazine, find pictures of things that are needs, cut them out and paste them on a piece of paper.

-Are there people in your local community that can't afford their basic needs? Talk with your teacher/parent and find one way you can help them.

-Complete the Wants Versus Needs worksheet in your activity book.

Project

It Costs How Much?

Do you know how much basic needs cost? To give you a real world idea of the cost of needs, you are going to interview adults on the basic costs of essential needs. Using index cards, compile a list of the following needs and their cost. Use a different index card for each item. Write one cost on each line going down. Ask more than one adult.

Housing and its necessary costs (use a different index card for each item below)
Rent
Mortgage (home repair savings monthly, yearly taxes)
Electric
Water (and sewer if applicable)
Trash (if applicable)

Clothing
What is the average monthly (or yearly cost broken down monthly) for one person?
Food
What is the average monthly cost for a family?
Car/Transportation (use a different index card for each method of transportation given)
What are options for transportation? What is the average cost for each option? For a car, what are the costs associated with having a car? Oil changes, maintenance budget, car insurance, gas. What do these cost on average?

Using your index cards, add each cost given to you in the interviews. Once you've added them, find the average. You do this by dividing the sum you just added, by the number of things you added. For example, if you interviewed 3 people, you will have three things to add and then you divide that sum by the number 3 to find the average. Use these averages to type up or write one to two paragraphs about what you learned. Did you expect things to cost as much as they did? What surprised you the most

Educator's Corner:

During the discussion on needs, your students may say a cell phone is a need. This is an opportunity to discuss a need to communicate with others to facilitate our basic needs, however having the newest phone that just came out isn't a need, using it for games isn't a need, etc. You can also discuss other things we enjoy for comfort, but maybe aren't an immediate need such as a computer, pets, TV, etc.

For the project, you can also discuss if these average costs varied by place if the people they interviewed lived in various locations. Is rent more expensive in Atlanta, GA as opposed to a smaller town?

Skills Learned

- Group discussion
- Critical thinking
-Writing/typing skills
-Research skills
-Interviewing skills
-Art/creativity
-Note taking/organization
-Compassion for others
-Money goals
-Every day math concept (addition and averages)

Money Lit Lesson Plan: Month 6

OPPORTUNITY COST & PRODUCT COMPARISON

Recap: Last month we discussed things we need to have to survive, versus things we want to have and how we need to choose to spend our money on needs first and set goals for wants.

This month: We are discussing how choices are made, what opportunity cost is, and how to compare similar products so we spend our money wisely.

Key Terms

Opportunity Cost: When you have to decide between two different options or things, the one you do *not* choose is your opportunity cost (opportunity lost).

Product Comparison: Using facts, information, and research to decide which of two similar things you like more and is worth spending your money on.

Limited Resource: Something that we need that there is only so much of, such as clean water, clean air, trees, money.

Resource: Something that can be used for a purpose such as tools, or materials to make things.

Memory Check-

What are goods and services? *What is revenue?* *What is an entrepreneur?*

What is human capital? *What is profit?* *What are needs and wants?*

MONTH #6: OPPORTUNITY COST & PRODUCT COMPARISON

Discussion Questions

Write your answers on a separate piece of paper

-Life is all about choices. In order to make good choices regarding money, we have to have all the information and a process for making the best choice. Today we are going to talk about the tools we need to make these smart money choices.

-Why is it important to make the best choice with our money? Money is considered a limited resource.

-You may have heard that clean water and air are limited resources. What are some other limited resources (things the world could run out of) you can think of?

-Money is just like that. I know I don't have a money tree to go pick money off of when I need to go to the store, or get gas, or pay my bills. The only way to replace money you've spent is by working to replace it. After seeing last month how much things cost, you can see that we need quite a bit of money just to cover the basics. This doesn't even include the things we want. Do you remember that list of wants you made? Wouldn't it be nice to buy a few things off that list too?

So let's talk about opportunity cost. Opportunity cost is when you have to choose between two different things, do I want to spend my only $10 on seeing a movie or buying a toy, and the thing you decide not to do is opportunity cost. It is the opportunity you had to spend your money, that you did not go with. For example, if you chose to go to the movie, buying a toy would have been your opportunity cost. If you had chosen to buy the toy, seeing the movie would have been your opportunity cost.

Life is full of choices! So how do we know which is the best choice to make? We have to think through the things that mean the most to us when making a decision. Is there just one kind of mac n cheese? Just one kind of cracker or cookie? One kind of candy? No! There are so many choices to make in life even at the grocery store. So since our money is limited, we have to decide what the best products (goods) to buy are.

On the next page you will find a useful process to use when making any decision.

The **PACED** decision making process can help.
-State the **P**roblem
-List **A**lternatives
-Identify **C**riteria
-**E**valuate alternatives based on our criteria
-Make a **D**ecision

Let's use our example above to better understand how to use the PACED process to make a decision.

1. Problem: I only have $10 and I want to go see a movie and there is a toy I want to buy.
   a. Alternatives: Go to a movie or buy a toy
   b. Criteria: If I go to a movie I will get to spend time with a friend or our family so it's an experience that would be fun. I also want the toy and if I buy it, I can play with it more than the one time we could see the movie.
   c. Evaluate: Which is more important, having a one-time experience (but maybe memory for a lifetime) or buying a toy that I could play with more than once (but could be lost or stolen or left somewhere).
   d. Decision: Based on what is important to me, I believe I will choose the toy because I can play with it more than once and can invite a friend over to play with it too.
2. What's a decision you have to make today or this week?
   a. What are your alternatives:
   b. What are the criteria:
   c. Evaluate your options:
   d. Make a decision:

We can use this process when we buy things we want and need. Did you know that between boxed, frozen, and restaurant styles, there are at least 43 brands of mac n cheese out there? So how do we decide which one is the best? How will we decide what car to buy? What computer to buy? What gaming console is best? Anytime you spend your limited resource you've earned (money), you should compare brands, styles, and prices to make sure you are spending wisely. The PACED decision making process can be useful for decisions like these as well.

Activities

-Pull out your list of wants and use the PACED method to decide between two of the items on your list. Once you've decided on two, set a money goal to save to buy them. Write a paragraph explaining why you made this choice.

-Complete the Opportunity Cost worksheet in the Money Lit activity book.

-Complete the Choices at the Skating Rink page in the Money Lit activity book.

-Cut out and paste the bookmark on the PACED method. Use in your favorite book to remind you to make good decisions!

Project

Product Comparison

Our project this month is about comparing products. Do you have a favorite mac n cheese brand? Fruit snacks? Yogurt brand?

1. Choose two items that you know you love a specific brand of. For example a specific brand of mac n cheese that is your favorite, or your favorite brand of cereal.
2. On a poster board, list your two items, in two columns. List their brand, and why you prefer them (texture, taste, color, etc).
3. Next time you go grocery shopping, choose a different brand, other than your usual brand, of those two items to try. For example, if you only ever buy Annie's mac n cheese, buy Kraft or Velveeta.
4. Your mission this month, is to try this other brand (it's good to try new things!) and then list facts about this product. Was the taste better, texture better, portion size it made better? Or was it still not your favorite and there were enough criteria about your usual brand that you like better?
5. On your poster board, list the qualities of the other brand you tried. Compare the two and decide which one is better. Then present your decisions and choices to your teacher.

Educator's Corner:

During the discussion on the PACED method, personalize some decisions they have to make and follow through the steps with them. Practicing this method will help it become second nature to them when making a decision, big or small. You can even use it on something as small as buying a pack of gum. It will teach critical skills thinking on an everyday level.

For the project, you can also add a math element by comparing prices of brands.

Skills Learned

- Group discussion
- Critical thinking
-Writing/typing skills
-Research skills
-Product Comparison
-Art/creativity
-Attention to detail
-Goal setting
-Trying new things
-Decision making skills
-Everyday math (price comparison)
-Self-esteem from good decision making
-Personal Independence

Money Lit Lesson Plan: Month 7

GOAL SETTING

Recap: Last month we discussed how to make wise money choices and how to use the PACED method to make many life decisions.

This month: We are discussing the most important tool for life, setting goals. This can apply to getting homework done, getting tasks done at work, accomplishing personal goals, and budgeting our money.

Key Terms

Goal: Something you want to do that you plan steps to achieve. It can be personal, for school, work, or financial like saving for something you really want.

Short-term Goal: A goal you can set and work towards achieving in a year or less.

Long-term Goal: A goal you can set and work towards achieving that will take you longer than a year. This could be saving for a car, or meeting goals for the college you want to attend.

Memory Check-

What are goods and services? *What is revenue?* *What is product comparison?*
What is human capital? *What is opportunity cost?*

MONTH #7: GOAL SETTING

Discussion Questions

Write your answers on a separate piece of paper

-If you have never been somewhere before, can you just get in the car and drive there? It's very hard to get somewhere if you don't know where you are going. Setting goals is very much like this. When you set a goal for either a short period of time or a longer period of time, you are mapping out where you are headed in life.

-A short term goal is a goal that you set to accomplish within a year. An example of this would be saving for a toy, cell phone, going to the movies, spending money on vacation, getting an A on your next test, or speaking kindly to someone who isn't always kind to you. These are activities that can be accomplished within a short amount of time or less than one year.

-A long term goal is a goal you set that can take you anywhere from 1-5 years to accomplish. An example of this would be saving for your first car, setting aside money for college, saving to take a cruise around the world, learning a language, keeping up a certain grade average over time to get into a college you prefer. These are all goals that will require more than one year to work towards, but are just as important as short term goals. For these goals you will need to write down what steps it will take to reach your overall goal, and then set smaller short term goals that eventually lead up to this larger goal further down the road.

-When setting a goal you want to make sure it is something that you can actually accomplish and is something you care about sacrificing for. Do you remember the concept of opportunity cost? When you are saving for something you will most likely need to practice delayed gratification, waiting longer for something you want instead of having other things now. When you are working towards a goal, you will have to delay the things you want and choose some things over others.

-For example, if you say you want to save for a toy that costs $15 and you want to buy it in 3 weeks. If you only make $5 a week in allowance or from jobs around the house, you will have to give up spending that $5 on candy or any smaller toy that would bring you instant joy so you can save your $15 in three weeks to buy the bigger item you wanted.

The candy and other smaller things you give up during those three weeks while you were saving money to reach your goal was your opportunity cost.

You want to make sure your goal is a SMART goal. What is a SMART goal?

Specific: I want to buy a LEGO set.
Measurable: To buy a LEGO set, I have to save $50.
Attainable: I receive $10 a week in allowance so I know I will have money to reach this goal.
Rewarding: Make sure your reward at the end is something you are motivated to save for.
Time: I can save for this in 5 weeks if I save all of my $10 each week.

Activities

-On a piece of paper, make a column for short-term goals, and a column for long-term goals. Under the short-term column, write down all the goals you have for the next week, month, year (get an A on my test, save for an item, score a goal). Repeat this process for your long-term goals (buy a car, go to a specific college). Save this list somewhere you can see it. As you complete your goals, you can check them off. Each goal you complete makes you a more skilled, smarter person.

Project
Accomplish Smart Goals

Let's practice this SMART goal setting process.

1. Choose one of your short-term goals, and one of your long-term goals from the list you made during your activity.
2. On a poster board, piece of paper, or you may type it, list your short-term goal. Give yourself some space on the page and then list your long-term goal.
3. List the steps it will take you to reach your short term goal.
4. List the steps it will take to reach your long-term goal. There will probably be more steps for this one. It helps to break long-term goals, or even big projects, down into smaller steps that are easier to accomplish.
5. As you accomplish these steps, celebrate the small wins that lead to you bigger win of reaching your goal!

Educator's Corner:

This process of goal setting is great for encouraging personal responsibility. This process of written goals will help your student to one day be able to set a written budget. A budget is just a set of goals for our money.

For a lesson in self-esteem, celebrate in a small way each time they accomplish a step. This encourages the positive behavior so it will continue. It can be as small as verbally acknowledging that they completed it.

Skills Learned

- Group discussion
- Critical thinking
-Writing/typing skills
-Attention to detail
-Goal setting
-Self-esteem building
-Personal accountability
-Decision making skills
-Stepping stone to the budgeting process

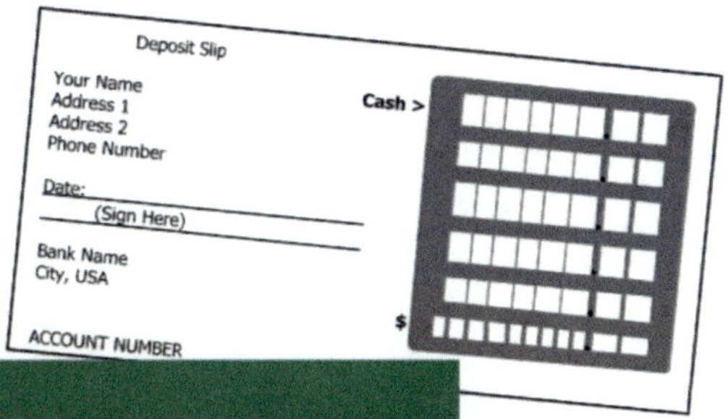

Money Lit Lesson Plan: Month 8

HOW SHOULD I PAY?

Recap: Last month we discussed goal setting and how to set SMART goals to achieve either short-term or long-term. Do you remember how long it should take to meet a short-term goal?

This month: We are discussing what the best ways are to pay for things we want and need.

Key Terms

Debit Cards: A card give to you by your bank when you open a checking or savings account. It is directly linked to the money you earn and deposit into your account.

Credit Card: A card you apply for from a credit card company that is money you are borrowing from your future self to spend now. You will pay interest on top of the amount you borrow.

Interest: A percentage determined by the bank or credit card company that you will owe based on how much you borrowed.

Memory Check-

What is the PACED method? *What are needs?* *What is a SMART goal?*
What is human capital? *What is opportunity cost?*

MONTH #8: HOW SHOULD I PAY?

Discussion Questions

Write your answers on a separate piece of paper

-Now that you have learned about setting goals and making smart decisions with your money, let's talk about how you should pay for things to best stay on track. First, you must know the various types of expenses you could have. Our expenses go on a *budget* which is a way to track your expenses or potential expenses.

-You can write your budget down on paper, can keep up with it on a spreadsheet, or in an app on your phone. A budget is a tool you use to plan out how to spend the money you make for the month. You write in what are called fixed and variable expenses. These expenses are usually planned, but it is good to factor in some money into your budget for things you might not have planned for.

There are fixed expenses which are things you pay the same amount for each month such as a house mortgage or car payment. When you buy a car, for example, there is a total cost of the car. Let's say a car is $10,000 total. The car salesman will take that total and divide it by the number of months you will pay for and add on interest for the loan. When this is done, your monthly payments remain the same each month ($250 for example) and you will pay that same amount monthly. It is a fixed cost and the amount won't change.

-Variable expenses are cost amounts that do change each month and can vary. A good example of this could be the amount you pay for groceries, gas in the car, or your electric bill. Your electric bill is based on how much of it you use and the amount will change each month based on that usage. That makes these costs variable expenses.

-Planned expense is something you knew you would have and put into your budget. An example of this would be rent or mortgage, electric and water expense, a car payment, food and gas.

-Unplanned expenses are things that come up suddenly such as you car breaking down, a flat tire, you get sick and have to go to the doctor, your air conditioning breaks and you have to call a repair person.

-So now that we know what types of expenses we spend our money on, let's talk about how we should pay for them. What types of ways have you paid for things you bought? Have you watched your parents pay for things in a different way than you have? You may have paid with cash. You may see that your parents pay for things with both cash and cards. Do you know what types of cards they are using?

-Debit cards are cards that you can swipe to pay for expenses. When you pay with a debit card, you are using money that you or your employer have put into your bank account. Instead of driving to the bank and taking cash out, you are using this debit card instead to make buying your items more convenient.

-Credit cards are cards that you can also swipe to pay for expenses. They are different from debit cards because when you use these, you are borrowing money to pay for your purchases. You are borrowing from money you haven't earned yet so in a sense, you are borrowing from your future self.

-You have to apply to use a credit card and they tell you, based on how trustworthy they think you are, how much you can borrow on the card. If you have paid your previous loans back on time they are more likely to approve for you to borrow more money than if you haven't paid on time (not trustworthy to pay a loan back on time). If you have never borrowed before at all (they have no idea if you are trustworthy because you haven't had the opportunity to prove you are).

-Since you are borrowing money when you use a credit card, you have to pay back what you spent plus interest. Interest is what the bank or credit card company earns from letting you borrow the money.

Activities

-Using the Debit and Credit Card pages in the Money Lit activity book as a guide, decide which payment method to use for the following situations. Write your answers on a piece of paper.

*Buying groceries at the store. *Buying a pack of gum. *Buying a plane ticket.

*Renting a car. *Buying a movie ticket. *Going to a concert. *Buying gas.

*Buying Christmas presents. *Buying a drink at a gas station. *Buying a computer.

*Going on vacation. *Paying your electric bill. *Buying snacks at a concession stand.

Project

Credit Card Interest Fees

Today we are going to use some everyday math to see the cost of using credit to pay for things. It's not bad to use credit to buy things, but once we know how credit works, we can make wise choices with how and when we use it.

The most commonly used credit cards are: Discover, American Express, Capital One, Barclays, Citibank, Bank of America, and Chase.

1. Write down or type those credit cards with space next to each one for some information about them.
2. Look up (research) each of these credit cards. They typically offer different types of credit cards. On some of the websites you can click on the "all cards" tab and it will show you the proposed benefits of each one. Some don't have fees, some have a lower APR (annual percentage rate - the amount of interest they charge each year, broken down over 12 months, to use their card).
3. Choose any card and click on "learn more" or similar tab that will give you information about that card. You are looking for the APR they charge to use the card. It will usually give a varying percentage, this is because the APR is based on your credit score.
    a. Your credit score shows them how trustworthy you are with money. Your score is higher the more credit you have used and paid back consistently on time. When you turn 18, you will have zero credit and should you choose to take out a loan to pay for your car, house, college, you will need to build trust with the bank.
    b. Take the two varying numbers they give you for the APR and find the average.
4. Now that you have an average APR interest fee amount, you can divide it by 12 months to find the monthly fee they charge This interest fee amount is multiplied by your credit card balance (the amount you've spent over that month) and that will be the final amount you owe for that month.
    a. For example, if you spent $100 that month and the interest fee is 16% it would look like this: $100 x 16%= $116. So you owe the $100 you spent plus the $16 they charged you in interest for a total of $116.
    b. If you choose not to pay that balance that month, the next month your beginning balance is the $116 amount. Let's say you spend another $100 that month. Credit cards work on compounding interest so the equation would look like this: $116+$100=$216 (ending balance). $216 x 16%= $34.56. Then you add your interest to your monthly balance to find the amount you owe for month two: $216+$34.56=$250.56.
    c. So after spending $200 over two months, you don't just owe that $200, you now owe $250.56. That $50.56 is the interest fee. You wouldn't pay this fee if you had paid the $100 at the end of month one, instead of letting it carry over. As long as you pay the balance each month, you won't incur fees.
5. Find this APR for five major credit card companies. Look them up, choose a card, find the average APR fee for one card, assume you spend $100 each month, and find the cost for both months with the understanding that the interest is compounding in month two.

Educator's Corner:

A lesson you could pull from the compounding interest cost of using credit cards, you can also explain compounding interest in a positive way and explain how it works when we invest money. This topic will be covered more in the investing chapter, but you could use this opportunity to bridge the learning gap of how compounding interest can work in the other direction as well. So if they were to take that amount that would have been charged in month one if the balance had rolled over ($16), and invested it and let it compound your money would work for you and you would have made money instead of owed it.

Skills Learned

- Group discussion
- Critical thinking
- Writing/typing skills
- Attention to detail
- Compounding interest
- APR interest
- Informed spending
- Decision making skills
- Everyday math

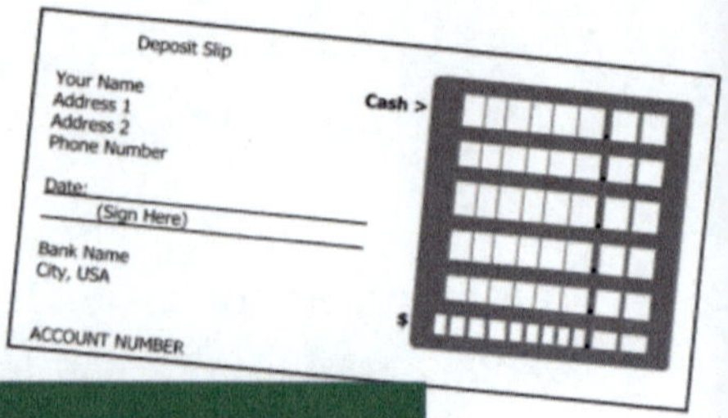

Money Lit Lesson Plan: Month 9

SAVE YOUR MONEY

Recap: Last month we discussed how to pay for things we want and need so we can make smart choices with our limited money resource.

This month: We are discussing how and why we should save some of the money we earn.

Key Terms

Savings: Money we set aside each time we earn money so we can meet our money goals and have money for later when we can't earn it.

Debt: Money you owe someone after borrowing it. You can be in debt when you borrow money from a friend, family member, the bank, a car dealership, or a credit card company.

Emergency Fund: Money you save in an account for when you can't work or when an emergency happens like a flat tire, your house needing a roof, or a medical emergency.

Memory Check-

What is the PACED method? *What are needs?* *What is a SMART goal?*
What is interest? *What is opportunity cost?*

MONTH #9: SAVE YOUR MONEY

Discussion Questions

Write your answers on a separate piece of paper

-So far you've learned;

- Why it's important to stay in school to continue learning and investing in yourself and your skills.
- We need to earn money to buy goods and services, that provide for our basic needs and wants.
- How to make wise spending decisions using the PACED method.
- How to set money goals so we can afford the things we want in addition to the things we need.
- How to pay for your goods and services.

Now it's time to talk about how to save your money and why that's important. "Sometimes it's not about how much money you make, but about how much money you keep, and how hard it works for you…". This is a quote by Robert Kiyosaki who wrote a book about what he learned from the adults around him on how to make, save, and invest money. His wise advice is that it's good to make money, but we don't always have to be millionaires to be doing well. If you know how to keep money in your pocket once you earn it, you will be better off than most of the other people you know (unless you share this secret with them too, which you should!). What do you like to do with money when you earn it?

A survey asked people how they spend their money and if they owe anyone money, 80 percent of people in America said they are in debt. Debt means you borrowed money either on a credit card or asked the bank to give you money in the form of a loan. When you do this you are borrowing from your future self, which means you are spending money you haven't yet earned.

When it says 80 percent of Americans are in debt, it means that a lot of people have borrowed money to pay for things they can't yet afford, but might be able to afford if they can pay it off in the future. Most people will be in debt to borrow money for big purchases like cars, houses, and sometimes college. Borrowing money for these expensive things is hard to avoid, but there are ways to help lower that cost such as applying for scholarships to help pay for college, or saving and investing money to help pay a larger down payment on your car or house. We should not go into debt to buy things like food, gas, furniture, toys, or vacations. Remember the concept of compounding interest? It applies to loans too. The larger the down payment, the less interest that compounds.

-If you can learn to wait, and delay instant gratification, to buy things until you have earned the money, you will have more money than most of the people around you because you learned that saving money is better than borrowing money.

-Now that we know why saving is so important, let's learn how to save. When you're an adult, it is recommended that you save at least 15 percent of each paycheck you earn. You can consider it as paying yourself money, before you pay all of your bills.

-When you make money now, you should save some, spend some, and give some. How much you save is up to you and your parents, but it is important to get used to saving money and paying yourself first. How much of your allowance or earnings do you usually save?

-We need to save some of our money to pay for things in the future we can't afford now, such as a car, college, cell phone, lap top, vacation spending money, gaming console, or whatever else you set as your money goal.

-We also need saved money for things like emergencies. When you buy your first car, there might be emergencies such as breaking down, a dead battery, or a flat tire. It's a good idea to have money saved specifically in an emergency fund so you can have money to have it towed, repaired, or to get a new tire. Maybe you bought a new phone, but it dropped and you need saved money to pay to have the screen fixed. Your parents won't always be around to help you pay for things like this so it's good to get into the habit of saving for emergencies now.

-When you have a car and are paying for expenses for yourself, it's a good idea to have anywhere from $500 to $1,000 saved in an account separately from your other savings. This should be an account you don't touch. That sounds like a lot of money doesn't it? Hopefully by now, you've learned that things can be expensive and we need to be prepared so we can avoid unnecessary debt.

Activities

-Complete the Save Your Money activity in the activity book

-Read the "Ways to Save" page in the activity book.

-Cut out the "Savings Goals Bookmark" in the activity book. Use it to set a savings goal. Identify if this is a long-term or short-term savings goal. Keep track of your savings as you earn money.

-You may complete the "Design a Piggy Bank" page in the activity book, or you may buy or create a piggy bank for your savings.

▼ Project

Savings Accounts

Today we are going to learn to open a savings account. First we need to do some research so we can make an informed decision.

There are three different types of banks you can use to open a savings account: A credit union, a traditional bank, or an online bank. You are going to research the savings account benefits of each and make a decision on which one is right for you.

1. Type or write the three types of banks, giving space in between each for your research.
   a. Under each type of bank, list the following:
      i. Minimum money amount needed to open a savings account
      ii. Fees for savings accounts
      iii. Interest earned for savings accounts
2. Do an internet search for your local credit unions. You will type in credit unions in ____ (your town name). Find the tab for savings accounts and find the three pieces of information listed below. Write down, or type up, each as you go a long.
   a. Repeat this process for a traditional bank (Bank of America, Regions, or something similar), and then again for an online only bank (Ally, or something similar).
3. Once you have this information on each type of savings account, use the PACED decision method to decide which savings account has the best benefit for you.
   a. Things to keep in mind. It helps to go with a bank that allows online banking so you can have an app and keep up with your savings. Grandma sends you a check for Christmas? No problem, you can deposit it using your banking app.
4. Once you've decided, work with your parents to open your savings account and begin your savings journey to reach your money goals.

Pro tip: Most banks charge you to send you monthly paper statements. You can avoid this by signing up for electronic statements. All you need is an email address. When you receive these electronic statements, it's a good idea to look them over and make sure there aren't any charges or fees you didn't sign up for or pay for. You can also keep up with this on your banking app. Being hands on with your money is the smartest thing you can do for your future!

Educator's Corner:

A field trip to your local bank to ask these questions in person may also be fun depending on the time you want to invest into this. You can pull out some everyday math with the percentage earned in savings accounts. If an online bank offers 3% a month interest, for example, you can have your student do the compounding interest math to see what that will amount to if they invest $50/mo with a compounding 3% interest. If you compare this to the other banking options, your student will see the value in choosing the right way to make their money work for them in this small way. They will see the value in both the interest given into the account, but also the value of adding money each month and watching the amount earned grow.

Skills Learned

- Group discussion
- Critical thinking
- Writing/typing skills
- Attention to detail
- Compounding interest
- Research skills
- Informed saving
- Percentages
- Decision making skills
- Debt versus setting savings goals
- The value of delayed gratification

Money Lit Lesson Plan: Month 10

INVESTING YOUR MONEY

Recap: Last month we discussed the importance of saving money and how to save money.

This month: We are discussing how and why we should invest some of the money we earn so it can work for us and grow.

Key Terms

Investing: Taking a percentage of our money and buying stocks, real estate, mutual funds, or buying a percentage of a company with the intention of that money making more money for us later.

Stocks: A stock is a portion (a share) of ownership in a company that they sell and you buy. This gives the company money to grow, and in turn your money should grow as well.

Stock Market: A money exchange market where you can buy and sell stocks or shares of a company.

Cryptocurrency: A digital coin built to only be spent or traded online.

Memory Check-

What is the PACED method? *Why do we need to save?* *What is a SMART goal?*
What is interest? *What is opportunity cost?*

MONTH #10:
INVESTING YOUR MONEY

Discussion Questions

Write your answers on a separate piece of paper

-This month we are going to focus on making our money work for us. How would you like make money even while your'e not working? It's possible. Does that mean you won't have to work? No. We still have to have a source of income to put into investing. But it does mean that we can take a percentage of our hard earned money and make it work for us so we don't have to work twice as hard.

-Investing is a process of taking a percentage of your money and putting it into the stock market, mutual funds, cryptocurrency, real estate, or new companies looking for investors. The goal with doing this is that your money will grow and gain compounding interest.

-In order for your money to grow, you need to do the research and make sure you are making an informed decision on where your money will go. Investing is taking a risk so you can lose your money. You could also grow your money by quite a bit. The goal is make the most informed decision you can.

-The most common investment is in the stock market. Apps like Stockpile and Robinhood make it easy to invest with a little bit of money so it's the easiest way to begin investing and to learn. If you are going to invest in the stock market, you need to know what you are putting your money in. This will require you to do some research into the companies you want to invest in first. Let's say for example you want to invest in Apple. If Apple is coming out with a new phone or other product soon, their stock might go up. If a company such as Sears is in the news for closing stores, their stock price is most likely going to go down, a lot. When companies do well, their stocks typically do too.

-Learning the ups and downs of the stock market takes practice. Most investments are long term investments meaning you want to choose a company that you think will do well and grow over time so your money will grow over time too. A good way to decide this is to think of what might be trending next. When Uber first started they didn't have Uber Eats. If you look up the Uber stock price it will show you the history over time of the stock going up and down. If you click on the five year view, you will see that in March of 2020, when the pandemic hit, the Uber stock went down because people didn't go anywhere or need rides. Thankfully they had Uber Eats as a part of their business model and you can see the steady growth of the stock as restaurants began offering food to go as a main way to stay open and keep business going.

-Thinking like that with each business will help you understand the rise and fall of stocks and feel more secure in your investment decisions. You always want to buy a stock when it's a lower price so your investment earns money as the price of the stock goes up based on how its doing in the economy. Sometimes stocks will go down because investors that have invested a lot of money sell their portion of the stock they bought. They do this to drive the price down, and then turn around and buy it again when the price is low so they can earn money as it goes back up.

-Cryptocurrency is another form of investing that is trending right now. Cryptocurrency is a form of money that was created digitally to be spent or traded online. You can transfer money from your bank to a digital coin bank and exchange your money for these digital coins. There are many types of digital coins.

-A few of the common ones are Bitcoin, Ethereum, and Litecoin. Each of these go up or down when people buy or sell them. Just like stock market, the price goes up and down as people sell and buy the digital coins. You can also make money by joining trading websites for cryptocurrency that trade cryptocurrency against other currencies.

-The person who invented cryptocurrency first invented Bitcoin. They only created so much of it, 21 million coins. Once people buy up all of it, there won't be anymore available unless someone sells theirs. When Bitcoin was first introduced in 2009, 1 Bitcoin was worth $0. Just like anything else in the economy, it needed demand for the price to rise. In 2011 it went from being worth $1 to $32 per Bitcoin. As of February 2021, 1 Bitcoin is worth $48,627.20. The price has gone up and down often over the years, and this past year it has been trending consistently upward, only going down when coins are sold.

-Unless you have $48,627.20, you will only be able to buy a portion of Bitcoin. Each $1 you spend buys you satoshis. A satoshi is the smallest unit of Bitcoin you can buy. It is similar to investing $5 into a larger stock like Tesla on a platform like Stockpile or Robinhood. Your money will still go up and down with the price of the stock or digital coin. What companies are you curious about investing in? Are you interested in investing in Cryptocurrency? Why or why not?

Activities

-Read over the "Cryptocurrency" page activity in the activity book

-Type up or write down some companies that you would like to own some shares of. What is trending right now? Will Snapchat continue to grow? Will Amazon keep growing as a company? Will Uber continue to offer services that they can bring to your door? Once you have a list of 4 or 5 companies you think you would like to buy a share of, look them up and look at their stock history. Save this information for next week's project.

Project
How Do I Invest?

- Using the "How Do I Invest In The Stock Market" page in your activity book, write down the companies you looked up last week and would like to invest in. In addition to those, also look up the following stocks:

Apple (AAPL), Tesla (TSLA), Nike (NKE), Snapchat (SNAP), Square (SQ), Netflix (NFLX)

Use the worksheet to down the ticker symbol (the abbreviation of the stock, for ex Apple is AAPL), and the price of the stock as it stands today for both the companies you looked up and the companies given above.

Every week, look up each stocks you wrote down and see how much it's gone up or down. Do an internet search of that company each week to track why their stocks may be going up or down. For example, search "Apple stocks rising May 2021" and see what other investors are saying or why they may be in the news in both positive and negative ways that are affecting their stocks.
At the end of four weeks, write in the price of each stock and compare your loss or gain for each. Buying multiple stocks is important to earning money. If one goes down, another grow and you will still earn money in that way.

At the end of the week and month, subtract the gains from the starting cost or the loss from the starting cost, to see how much it went up or down in value. If you want to, you can apply an invested value to each stock you tracked, say $5 each, and see how much money you would have lost or gained over the course of that week and month.

Once you have done the appropriate research and feel more confident and informed of how stocks going up and down. Speak with your parents about opening an investing account. Several good accounts are: Stockpile, Robinhood, or Ally Investing. They allow you to invest small portions of larger stock. If you don't have $129.84 to invest in one Apple stock, you can invest $5 to buy a portion of a whole stock. Things to look for when choosing a platform, the cost of each trade (buying and selling) and the minimums needed to invest.

Educator's Corner:

There are so many everyday math problems you could pull out of this month if you wanted to. You could set up a scenario with compounding interest over time to see how much they would earn over 1 year, 5 years, or 10 years of investing. It is important to note that most investing should be long term so they should be looking for companies that will grow over time. They can set a money goal for this investing money they are earning. For example it could pay for a first car, college, a down payment for their first home, a real estate investment or small business venture, or simply keep compunding for retirement. You can also do a fieldtrip with an investor to learn more about mutual funds, CD accounts, and other forms of long-term investing.

Skills Learned

- Group discussion
- Critical thinking
-Writing/typing skills
-Attention to detail
-Compounding interest
-Research skills
-Informed decision making
-Percentages
-Cryptocurrency
-Investing
-Long-term goals

Made in United States
Orlando, FL
04 November 2026

90200305R00029